ABC
Sheep Book

AO PRESS

Jessica Lee Anderson

Paperback ISBN: 979-8-9899560-7-4

To Ava, my sheep loving daughter—I love you more than there are stars in the sky and fibers of fleece in this world! - JLA

Photo credits—Front Cover: pkanchana, daseaford, neilkendall, mygtree, dabjola; Back Cover: Life On White; Cover Page: pkanchana, daseaford, neilkendall, herbert25120; Copyright Page: mygtree, Aumsama (Dorper), Irina Babina (Bighorn Sheep); Dedication Page: panaramka; p. 4: aksphoto; p. 5: alexeys; p. 6: rcimages; p. 7: MollyNZ; p. 8: namchetolukla; p. 9: Wirestock; p. 10: sarahdoow; p. 11: compuinfoto; p. 12: subtik; p. 13: Musat; p. 14: Michele Ursi; p. 15: daseaford; p. 16: Byrdyak; p. 17: Byrdyak; p. 18: silkfactory; p. 19: Wirestock; p. 20: caichuqing; p. 21: slowmotiongli; p. 22: Wirestock; p. 23: Wirestock; p. 24: Ondrej; p. 25: bloodua; p. 26: Wirestock; p. 27: PaitedLens; p. 28: DavorLovincic; p. 29: cmtyers; p. 30: Life On White (lamb), Frank Manzo (Wiltshire Horn Sheep) Life On White (Suffolk); p. 31: Michael Anderson

This Book Belongs to:

_____

 is for Awassi Sheep

Awassi Sheep are popular in Southwest Asia. The females, called ewes, produce milk, and the males, known as rams, have long, curled horns.

Aa

is for Babydoll Southdown Sheep

Babydoll Southdown Sheep are an old breed of English sheep. They are small in size, produce wool, and are called "organic weeders" as they graze on weeds but don't damage shrubs or vineyards.

 is for Cotswold Sheep

Cotswold Sheep are a large, tall breed of sheep from England. Their fleece is long with wavy curls.

Cc

D is for Dorset Sheep

Dorset Sheep are a white sheep breed with long bodies and wide faces. Dorset rams can have horns or they can be polled (no horns.)

 is Exmoor Horn Sheep

Exmoor Horn Sheep come from the hills of Exmoor National Park (United Kingdom). Both male and female Exmoor Horn Sheep can have horns.

Ee

F is for Finnish Landrace Sheep

Finnish Landrace Sheep are also known as Finnsheep. These sheep are native to Finland and produce many babies.

is for Greyface Dartmoor Sheep

Greyface Dartmoor Sheep are medium sized with short, woolly legs, and they don't have horns. They produce a good amount of wool, and they even have woolly faces!

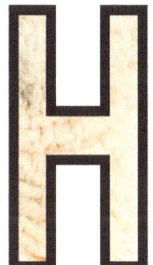is for Hampshire Down Sheep

Hampshire Down Sheep have dark faces, ears, and legs, and they have no horns. Their wool is white and downy, and it covers the tops of their heads and around their eyes.

Hh

I is for Icelandic Sheep

Icelandic Sheep are medium sized, hardy sheep from Iceland. Their long wool keeps them warm in the cold—they have an inner coat and an outer coat.

 is for Jacob Sheep

Jacob Sheep are small, spotted sheep that look a bit like goats. Both ewes and rams have horns, though the males have larger horns— they can have two, four, or even six horns!

Jj

K is for Kerry Hill Sheep

Kerry Hill Sheep have black noses as well as black and white markings on their faces and legs. These sheep have a soft wool, and they don't have horns.

K k

L is for Lincoln Sheep

Lincoln Sheep are sometimes called Lincoln Longwool Sheep—their wool grows up to twelve inches a year! They are one of the largest breeds of sheep.

L l

M is for Mouflon Sheep

Mouflon Sheep are a type of wild sheep originally from an area near the Caspian Sea. The rams have large horns which helps determine their rank in the herd.

Mm

 **is for Navajo-Churro Sheep**

![Navajo-Churro Sheep ram with multiple horns]

Navajo-Churro Sheep rams can have multiple horns. Their fleece has two coats, a soft to medium inner layer and an outer layer that is longer and coarser.

N n

 is for Old Norwegian Sheep

Old Norwegian Sheep are originally from Norway and the surrounding areas. Their fleece is fine, and the outer coat has long fibers.

P is for Pomeranian Coarsewool Sheep

Pomeranian Coarsewool Sheep are sometimes simply called Pomeranians, named for the region they come from. They have black heads with gray or slate-blue wool, and their lambs are born completely black.

is for Qinghai Black Tibetan Sheep

Qinghai Black Tibetan Sheep are on old breed of sheep from China. They live high on the Qinghai Tibetan Plateau.

Qq

is for Romanov Sheep

Romanov Sheep are named after the royal Russian Romanov family. They can survive tough winter conditions, and the ewes have the ability to give birth to multiple babies instead of just one or two like many other sheep breeds.

R r

21

is for Shropshire Sheep

Shropshire Sheep, originally from England, are medium to large sized sheep with dark faces. They are known for being woolly from nose to toes!

T is for Texas Dall Sheep

Texas Dall Sheep are not related to Alaskan Dall Sheep—they got their start in Texas as a cross between a domestic sheep and a wild sheep. Males and females have horns, though the males' horns can grow larger and more curved.

 U is for Urial Sheep

Urial Sheep are similar to Mouflons, and they are also known as Arkars or Shapo. They live in the mountains in South Asia.

V is for Valais Blacknose Sheep

Valais Blacknose Sheep are from Switzerland. They have woolly coats and black patches on their noses, eyes, ears, knees, and feet.

W is for Whitefaced Woodland Sheep

Whitefaced Woodland Sheep are hardy sheep with white wool from the British hills. Both rams and ewes have horns, and the rams can grow horns with spirals.

W w

X is for Xinjiang Finewool Sheep

Xinjiang Finewool Sheep were some of the first breeds of sheep in China. These sheep were originally crossed with Merino breeds, and they are known for their wool.

X x

Y is for Yemeni Sheep

Yemeni Sheep are found in Yemen. These sheep have fat tails (like a camel's hump that helps get the animal through hard times), and they are often "earless."

Z is for Zwartbles Sheep

Zwartbles Sheep are from the Netherlands. They have either brown or black fleece with a white stripe (or "blaze") on their face as well as white "socks."

Z z

5 Sheep Facts:

 Sheep have pupils shaped like rectangles that allow them to have a wide field of vision—this is important for them to stay aware of predators.

 Sheep are social creatures! A group of sheep is called a flock.

 Sheep have a dental pad instead of sharp upper teeth — this helps them to gather up grass and plants.

 Sheep have an upper lip that is divided by a groove called a philtrum which allows them to get close to the ground to graze.

 Some types of sheep fleece will keep on growing forever and needs to be sheared to keep the sheep light and cool. Other sheep, like the Wiltshire Horn Sheep, have fleece that sheds naturally.

Jessica Lee Anderson is an award-winning author of over 75 books for young readers. Jessica lives near Austin, Texas with her daughter, Ava, and husband, Michael. They are volunteer farm animal feeders at a living history museum called Pioneer Farms. You can learn more about Jessica by visiting www.jessicaleeanderson.com.

Check out these other titles:

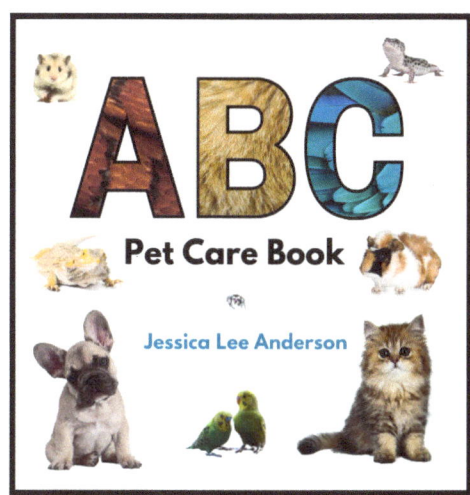

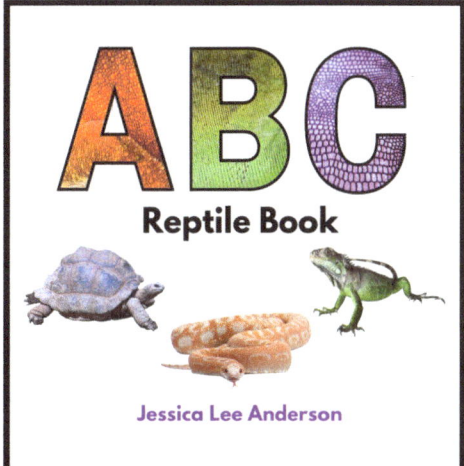

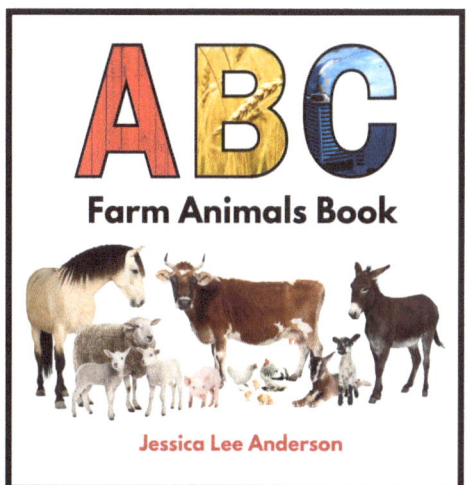

www.ingramcontent.com/pod-product-compliance
Lightning Source LLC
Chambersburg PA
CBHW041620120626
46551CB00003B/513